Using Expanded Notation to Represent Numbers in the Millions

Collierville Schools
146 College Street
Collierville, TN 38017

TARA OAKS ELEM.

Orli Zuravicky

PowerMath™

The Rosen Publishing Group's
PowerKids Press™
New York

Published in 2004 by The Rosen Publishing Group, Inc.
29 East 21st Street, New York, NY 10010

Copyright © 2004 by The Rosen Publishing Group, Inc.

All rights reserved. No part of this book may be reproduced in any form without permission in writing from the publisher, except by a reviewer.

Book Design: Ron A. Churley

Photo Credits: Cover © Michael S. Yamashita/Corbis; cover (left and right insets), pp. 10–11, 30–31 © Jim Zuckerman/Corbis; cover (bottom inset), pp. 4–5, 32 (fossil) © Stephen Silkes/The Image Bank; pp. 4–5, 32 (background) © Corbis; pp. 6–7, 9 (inset), 20 (inset) © PhotoDisc; pp. 8–9 © Richard Coomber/Taxi; p. 8 (inset) © J. Brian Alker/Taxi; pp. 12–13 © RO-MA Stock/Index Stock; p. 12 (inset) © John Warden/Index Stock; pp. 14–15 © Sally A. Morgan; Ecoscene/Corbis; pp. 17, 22–23 © J. Silver/SuperStock; p. 19 © Jonathon Blair/Corbis; pp. 20–21 © Sanford/Agliolo/Corbis; p. 22 (inset) © Gunter Marx Photography/Corbis; p. 24 (both) © Hulton/Archive; pp. 26–27 © Kevin Schafer/Corbis; pp. 28–29 © Richard Cummins/Corbis.

Library of Congress Cataloging-in-Publication Data

Zuravicky, Orli.
  When there were dinosaurs : using expanded notation to represent numbers in the millions / Orli Zuravicky.
      p. cm. — (PowerMath)
Summary: Introduces the concept of expanded notation by providing facts about dinosaurs and the periods in which they walked the earth millions of years ago.
  ISBN 0-8239-8988-7 (lib. bdg.)
  ISBN 0-8239-8901-1 (pbk.)
  6-pack ISBN: 0-8239-7429-4
  1. Decimal system—Juvenile literature. 2. Dinosaurs—Juvenile literature. [1. Decimal system. 2. Number systems. 3. Dinosaurs.] I. Title. II. Series.
  QA141.35. Z87 2004
  513.5'5—dc21

2003001008

Manufactured in the United States of America

Contents

Working with Numbers — 4

A Long Time Ago — 6

How Dinosaurs Came to Be — 9

Jurassic Giants — 12

Rulers of the Land — 16

Disappearing Act — 21

Dinosaur Discoveries — 25

Dinosaur Skeletons — 28

A World of Numbers — 30

Glossary — 31

Index — 32

Working with Numbers

For thousands of years, people have used numbers to help them understand their world. Even the earliest people had counting systems to help them figure out things like when to plant crops and when to **harvest** them. Numbers are still a huge part of everyday life. Numbers can tell us how long ago the first people lived and how far away Earth is from Mars. Numbers can tell us the time and the **temperature**. Numbers can tell us how much things cost in the store and the ages of our friends and family members.

Numbers can be tricky. Some are small fractions while others can be 9 or more **digits** long! Such big numbers might seem confusing at first, but we can break them down into smaller parts to understand them. This process is called **expanded notation**.

We usually write numbers in this form: 523,343,724. We would say this number as "Five hundred twenty-three million, three hundred forty-three thousand, seven hundred twenty-four." Expanded notation will show you how to break down this number and other numbers and look at every digit in terms of the place value it belongs to: hundred millions, ten millions, millions, hundred thousands, ten thousands, thousands, hundreds, tens, or ones.

Let's go through the place groups of this number: 523,343,724. Expanded notation means taking each number and writing it in its fullest form, showing each digit in terms of the place value it occupies in the number.

500,000,000	(5 hundred millions) - hundred millions place value
20,000,000	(2 ten millions) - ten millions place value
3,000,000	(3 millions) - one millions place value
300,000	(3 hundred thousands) - hundred thousands place value
40,000	(4 ten thousands) - ten thousands place value
3,000	(3 thousands) - one thousands place value
700	(7 hundreds) - hundreds place value
20	(2 tens) - tens place value
4	(4 ones) - ones place value

Since 5 is in the hundred millions place, we would write it as 500,000,000. Look at the box on this page to see how we would write the rest of this large number using expanded notation.

A Long Time Ago

Using expanded notation can help us understand any subject that uses large numbers, like the long periods of time in Earth's history. Let's go back in time millions of years, to when there were dinosaurs!

Dinosaurs first lived on Earth about 230,000,000 years ago. They remained on Earth for about 150,000,000 years, through 3 different scientific time periods. The earliest dinosaurs appeared during the Triassic (try-AA-sik) period. Many more appeared during the Jurassic (juh-RA-sik) period. Dinosaurs died out about 65,000,000 years ago, during the late Cretaceous (krih-TAY-shuhs) period.

During these different periods, Earth went through many changes. In the beginning, Earth's 7 continents were just 1 big piece of land. As time went on, the continents began to break apart and the climate began to change. As the world changed, the dinosaurs had to **evolve**, or change, with it. When they were no longer able to keep up with the changes taking place on Earth, they became **extinct**.

Let's break down some of these big numbers to understand how long ago dinosaurs actually lived. When writing a number in expanded notation, use addition and show the plus sign. When you add all the numbers together, you should get your original number, showing that you expanded the number correctly. Remember to work from the largest place value on the left side of the number to the smallest place value on the right side.

What is the expanded notation of 230,000,000 (two hundred thirty million)? We'll start with the "hundred millions" place value. How many hundred millions are there in this number? The answer is 2. Two hundred million is written 200,000,000. There is a 3 in the ten millions place. This is written 30,000,000. There are no more numbers in any of the other place values, so we only have 2 numbers to add: 200,000,000 and 30,000,000.

```
  200,000,000  (2 hundred millions)
+  30,000,000  (3 ten millions)
  -----------
  230,000,000
```

8

How Dinosaurs Came to Be

Where did dinosaurs come from? Scientists believe that dinosaurs evolved from reptiles that lived about 250,000,000 years ago! A reptile is a **cold-blooded** animal that lays eggs and has scales to protect its skin.

Earth's climate was hot and dry 250,000,000 years ago. As the climate changed, lakes and rivers dried up, leaving deserts and mountains where there had once been water. The animals that lived during this time had to **adapt** to these changes. Some animals developed body parts that helped them live on land rather than in the water. Over many years, some animals grew lungs to help them breathe air and legs to help them stand and walk on land. Between 230,000,000 and 220,000,000 years ago, these adaptations brought about the first dinosaurs.

Some people believe that animals like alligators and lizards evolved from dinosaurs!

TARA OAKS ELEM.

Since dinosaurs were living on Earth 220,000,000 years ago, let's expand the number 220,000,000. How many hundred millions are in this number? There are 2 hundred millions, which would be written as 200,000,000. What place is the next number 2 in? If you said the ten millions place, you're right! The written form for that number is 20,000,000. There are no more numbers in any of the place values, so now we can add 200,000,000 and 20,000,000 to get 220,000,000.

Another form of expanded notation uses multiplication as well as addition. In this form, you write 2 x 100,000,000 instead of 200,000,000. You'll learn more about this in the next few chapters.

```
  200,000,000   (2 hundred millions)
+  20,000,000   (2 ten millions)
  -----------
  220,000,000
```

Jurassic Giants

The Jurassic period was between 213,000,000 and 144,000,000 years ago. During the Jurassic period, the surface of the Earth was shifting. The 1 large piece of land—or "supercontinent"—was starting to break apart. Around this time, big plant-eating dinosaurs, called sauropods (SOHR-uh-pahdz), became more common. Sauropods are the biggest animals that have ever lived on Earth. They had four legs, long necks, small heads, and big bodies. Some stood over 50 feet tall!

Another common type of dinosaur of the Jurassic period was *Stegosaurus* (steh-guh-SOHR-us). *Stegosaurus* grew to be about 20 feet long and had giant plates running the entire length of its head and back. *Stegosaurus* was also a plant-eating dinosaur. Plant-eating animals are called **herbivores**. The biggest meat-eating dinosaur to live during the Jurassic period was *Allosaurus* (aa-luh-SOHR-us). Meat-eating animals are called **carnivores**.

Allosaurus

Stegosaurus

Herbivores like *Stegosaurus* had to be careful. Sometimes they were hunted and killed by meat-eating dinosaurs like *Allosaurus*.

It was also during the Jurassic period that the first known bird dinosaurs appeared. *Archaeopteryx* (ahr-kee-AHP-tuh-riks) was shaped like a bird and had feathers, but it also had a long bony tail, sharp teeth, and claws on its wings. Scientists believe that this bird dinosaur evolved from earlier small dinosaurs. In Germany, bird dinosaur fossils more than 145,000,000 years old have been found. Scientists also believe that these bird dinosaurs evolved into our present-day birds.

Besides using addition to expand numbers, we can also use multiplication. Let's use this method to expand the number 145,000,000. There is a 1 in the hundred millions place. Therefore, it can be written 1 x 100,000,000, which is the same as 100,000,000. The second group is the ten millions, of which there are 4. This becomes 4 x 10,000,000, which is the same as 40,000,000. The next number is a 5. Can you finish expanding this number?

Did you write 5 x 1,000,000? You're right! Five is in the one millions place, so multiplying it by 1,000,000 will give you 5,000,000.

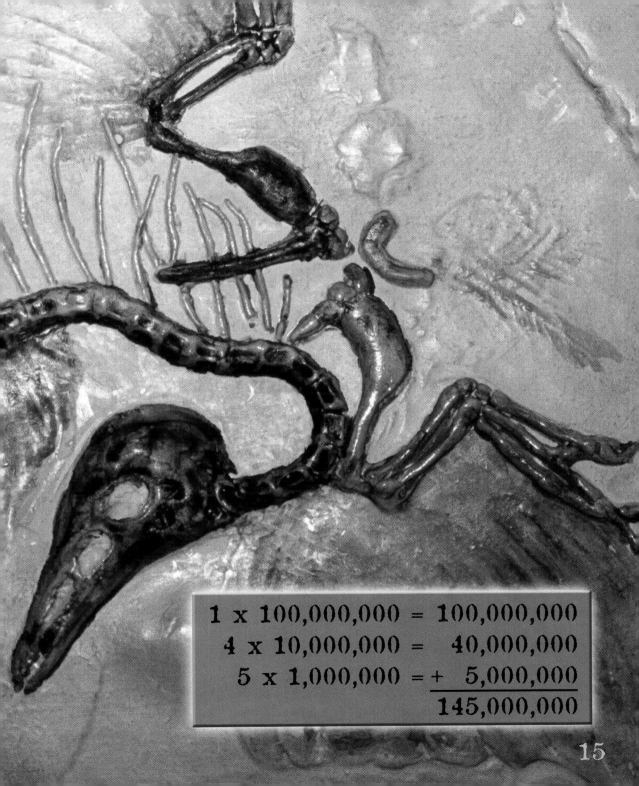

Rulers of the Land

By the beginning of the Cretaceous period, the continents as we know them were continuing to separate. An ocean—which we know today as the Atlantic Ocean—had already formed between North America and Africa, but North America and Europe were still attached in the north. New forms of plant life had begun to grow. Flowering plants, palm trees, oak trees, and willow trees all appeared as the climate and land changed.

During this time period, dinosaurs had taken over the land. The biggest meat-eating dinosaurs ever to walk on Earth appeared! The smartest and most powerful carnivore, *Tyrannosaurus rex* (tuh-ra-nuh-SOHR-us REKS), appeared in the late Cretaceous period, between 85,000,000 and 65,000,000 years ago.

Let's work with multiplication again to expand the number 65,000,000. There is no number in the hundred millions place this time, so you'll start with the ten millions place.

Tyrannosaurus rex was over 40 feet long. It stood and walked on its huge, powerful back legs. Its front legs were tiny and weak. *Tyrannosaurus rex* also had thick, sharp teeth used for tearing off meat and crushing bones.

Tyrannosaurus rex

```
6 x 10,000,000 =   60,000,000
5 x  1,000,000 = +  5,000,000
                   65,000,000
```

Even though *Tyrannosaurus rex* was strong and powerful, it was not the largest meat-eating dinosaur. The largest of all the carnivores was *Giganotosaurus* (jih-guh-not-uh-SOHR-us), who appeared on Earth about 95,000,000 years ago. *Giganotosaurus* could grow up to 50 feet long and weigh about 16,000 pounds. That's 8 tons!

New forms of herbivores also appeared around this time, including horned dinosaurs like *Triceratops* (try-SAIR-uh-tops), which lived in present-day western North America. The name "*Triceratops*" means "three-horned face." *Triceratops* had 3 horns on its head—1 over each eye and 1 over its nose. Its head alone could grow to be over 7 feet long!

Let's do the expanded notation for 95,000,000. The box on page 19 shows the number 95,000,000 in both forms of expanded notation. You can see that whichever way you do the expanded notation, the answers should always be the same!

Triceratops

```
9 x 10,000,000 =   90,000,000          90,000,000
5 x  1,000,000 = +  5,000,000   or   +  5,000,000
                   95,000,000          95,000,000
```

Disappearing Act

By the late Cretaceous period, the dinosaurs began to disappear, along with many other kinds of animals. No one really knows why this happened, but scientists have many ideas about why the dinosaurs became extinct.

Some scientists believe that a large **meteor** hit Earth, sending so much dust into the air that sunlight could not reach Earth. Without sunlight, plants died, causing plant-eating dinosaurs to die because they had nothing to eat. Meat-eating dinosaurs—who ate plant-eating dinosaurs—then died, too.

Other scientists believe that many volcanoes erupted all over Earth at that time. The volcanoes produced a lot of lava, which killed many life-forms, including the dinosaurs. Still other scientists believe that a change in climate created a world in which the dinosaurs could no longer live.

Some people believe that the meteor that hit Earth could have been as much as 10 miles across!

It is possible that a combination of these 3 events led to the end of the dinosaurs. Before they became extinct about 65,000,000 years ago, dinosaurs had lived on Earth between 150,000,000 and 165,000,000 years! Humans have only lived on Earth for about 1,000,000 years.

Let's pick a number between 150,000,000 and 165,000,000 and write it using the addition form of expanded notation. This time, let's choose a number that doesn't have so many zeros! How about 154,362,793? That should be a challenge!

$$\begin{array}{r}100{,}000{,}000\\50{,}000{,}000\\4{,}000{,}000\\300{,}000\\60{,}000\\2{,}000\\700\\90\\+\phantom{000{,}00}3\\\hline 154{,}362{,}793\end{array}$$

You can look at the box on page 5 to help you remember all the place values.

Dinosaur Discoveries

People have been finding dinosaur fossils for hundreds of years. However, people did not always know what the fossils were! A British professor named William Buckland was the first person to identify a dinosaur fossil and give it a name. In 1824, Buckland was shown the bones of a huge animal, including a lower jawbone that was filled with sharp teeth. He named the animal *Megalosaurus* (meh-guh-luh-SOHR-us), which means "giant lizard."

In 1842, a British scientist named Sir Richard Owen was the first person to use the term "dinosaur." "Dinosaur" is a Latin word that means "terrible lizard." In 1854, Owen and a man named Benjamin Waterhouse Hawkins helped to construct huge, life-sized models of what they thought *Megalosaurus* and other dinosaurs looked like. The models still can be seen today at Crystal Palace Park, just south of London, England.

The dinosaur models that were constructed for Crystal Palace Park were based on the fossils of the dinosaurs that had been found in England. These models gave many people their first look at the huge creatures that roamed Earth millions of years ago.

Since the discoveries of Buckland and Owen, people have continued to uncover dinosaur fossils, and scientists are still trying to piece together the history of the dinosaur age. Dinosaurs with feathers have recently been discovered in China! These dinosaurs lived more than 120,000,000 years ago.

Scientists have recently discovered dinosaur fossils over 235,000,000 years old in Brazil! These may be the oldest known dinosaur fossils found so far. The fossils date back to the middle of the Triassic period, when dinosaurs first appeared. Let's say scientists were to discover a fossil that was 245,589,300 years old. That fossil would be even older than the ones found in Brazil!

Let's get some more practice with expanded notation by working with the number 245,589,300. Can you use the multiplication form of expanded notation to write this number? Be careful of the numeral 5 in this expanded notation! It appears in 2 different place values.

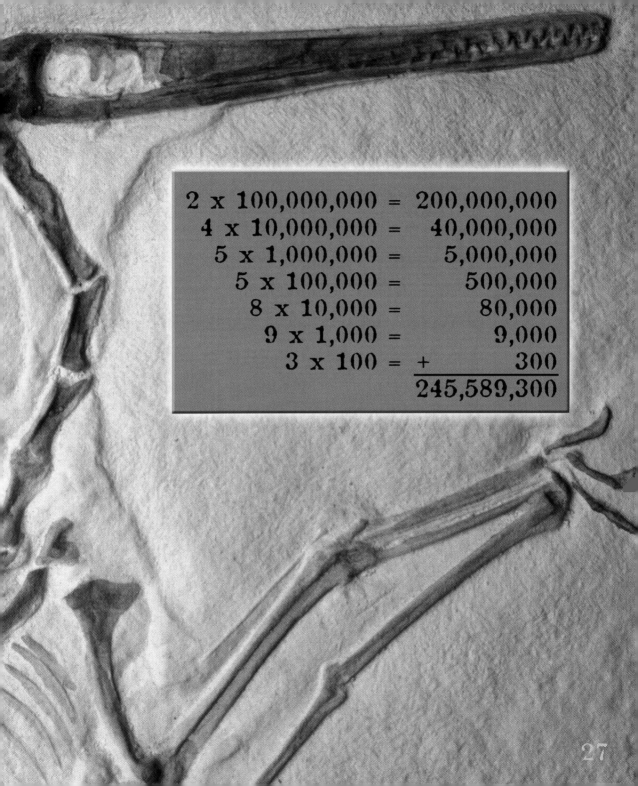

2 × 100,000,000	=	200,000,000
4 × 10,000,000	=	40,000,000
5 × 1,000,000	=	5,000,000
5 × 100,000	=	500,000
8 × 10,000	=	80,000
9 × 1,000	=	9,000
3 × 100	= +	300
		245,589,300

Dinosaur Skeletons

Between 1909 and 1912, scientists from Germany were searching for fossils in Tanzania, Africa, when they made an astonishing discovery—500,000 pounds of dinosaur bones buried deep in the dirt! The bones had been buried between 213,000,000 and 144,000,000 years ago, during the late Jurassic period. The bones belonged to a dinosaur called *Brachiosaurus* (bra-kee-uh-SOHR-us), one of the biggest dinosaurs to ever live. The scientists took the bones from several individual dinosaurs and fit them together to build the largest dinosaur skeleton ever seen. The skeleton is about 38 feet tall and 73 feet long!

Let's say some of these *Brachiosaurus* bones were 157,872,045 years old. Can you imagine dinosaur bones that old? Breaking down the number using expanded notation helps you understand a little better how long ago that was! Use your expanded notation skills to write out this number using addition.

```
    100,000,000
     50,000,000
      7,000,000
        800,000
         70,000
          2,000
            000
             40
  +           5
    157,872,045
```

Brachiosaurus was a sauropod, which was a plant-eating dinosaur. The skeleton of this huge dinosaur is on display at the Natural History Museum in Berlin, Germany.

A World of Numbers

Numbers teach us so many things about our world. If you don't understand how to read numbers, you could miss out on many fun facts! That's why knowing expanded notation can be really important. Now that you understand how to use it, not only can it help you to explore the amazing world of dinosaurs, but it can help you learn about many other things, too. If you keep practicing, soon you will become so familiar with numbers that you won't need to expand them anymore. You'll know what they mean just by looking at them!

Try to find other numbers for how long ago different dinosaurs lived or how old certain fossils are. Then practice writing the numbers in expanded notation.

Glossary

adapt (uh-DAPT) To change to fit new conditions.

carnivore (KAHR-nuh-vohr) A meat-eating animal.

cold-blooded (KOLD–BLUH-duhd) An animal whose body is as warm or as cold as the air around it.

digit (DIH-juht) Any of the figures 0, 1, 2, 3, 4, 5, 6, 7, 8, and 9.

evolve (ih-VAHLV) To change over a period of time.

expanded notation (ik-SPAN-duhd noh-TAY-shun) A way of writing out large numbers as an equation that lets you see which place value every digit belongs to.

extinct (ik-STINKT) No longer in existence.

harvest (HAHR-vuhst) To gather crops from the fields after they have finished growing.

herbivore (HER-buh-vohr) A plant-eating animal.

meteor (MEE-tee-uhr) A space rock that burns up when it enters Earth's atmosphere.

temperature (TEM-puhr-chur) How hot or cold something is.

Index

A
adapt, 9
adaptations, 9
Allosaurus, 12
Archaeopteryx, 14

B
bird dinosaur(s), 14
Brachiosaurus, 28
Buckland, William, 25, 26

C
carnivore(s), 12, 16, 18
Cretaceous period, 6, 16, 21
Crystal Palace Park, 25

E
evolve, 6
extinct, 6, 21, 22

G
Giganotosaurus, 18

H
Hawkins, Benjamin Waterhouse, 25
herbivores, 12, 18

J
Jurassic period, 6, 12, 14, 28

L
London, England, 25

M
meat-eating, 12, 16, 18, 21
Megalosaurus, 25

O
Owen, Sir Richard, 25, 26

P
plant-eating, 12, 21

R
reptile(s), 9

S
sauropods, 12
Stegosaurus, 12
supercontinent, 12

T
Triassic period, 6
Triceratops, 18
Tyrannosaurus rex, 16, 18